T0332403

"As the design of our homes becomes ever more sophisticated, the physics of those buildings becomes critical. Whereas our older building stock were to a certain extent tolerant of error, today's need for airtightness and insulation demand exacting standards. Whilst many professionals may have a working knowledge of dampness and its causes, it is vital that concepts of water transport, condensation and mycology are understood. This book brings together the skills and knowledge of established experts in the field and will provide a valuable link between scientific theory and its practical application".

—*Trevor Rushton FRICS FCABE, Technical Director, Watts Group Limited*

Dampness in Dwellings

This book provides a definition of dampness in each of its forms, details the various potential sources, and causes that can result in damage to the building, and importantly, the threats to the health of the occupiers. It is practical, providing an outline of the possible solutions looking at aspects of building design and construction that can reduce or avoid the risk of dampness. It also discusses why dampness is a risk to the health of occupiers and so justifies the need to protect health by reducing or removing it. Co-authored by a medical doctor and environmental health practitioners with a combined experience of over 50 years, this book includes:

- Explanations and justifications for why dampness is important, and why remedial action must be taken.
- Up-to-date information on the causes, effects, and remedies of damp in the housing environments.

Dampness in Dwellings is a pivotal resource for active professionals in the housing, medical, and legal sectors.

Professor David Ormandy is a public and environmental health advisor, housing and health consultant, researcher, and author based at the School of Law, University of Warwick, UK.

Dr Véronique Ezratty MD, is a medical doctor, environmental risk assessor, and author based in the Medical Studies Department of Electricité de France.

Dr Stephen Battersby MBE, is an environmental health practitioner, independent consultant, and advisor.

Routledge Focus on Environmental Health

Series Editor: Stephen Battersby, MBE PhD, FCIEH, FRSPH

For more information about this series, please visit:
https://www.routledge.com

Dampness in Dwellings
Causes and Effects

David Ormandy,
Véronique Ezratty, and
Stephen Battersby

Routledge
Taylor & Francis Group

LONDON AND NEW YORK

First published 2022
by Routledge
4 Park Square, Milton Park, Abingdon, Oxon OX14 4RN

and by Routledge
605 Third Avenue, New York, NY 10158

Routledge is an imprint of the Taylor & Francis Group, an informa business

British Library Cataloguing-in-Publication Data
A catalogue record for this book is available from the British Library

Library of Congress Cataloging-in-Publication Data
A catalogue record has been requested for this book

ISBN: 978-0-367-53039-6 (hbk)
ISBN: 978-0-367-53040-2 (pbk)
ISBN: 978-1-003-08023-7 (ebk)

DOI: 10.1201/9781003080237

Typeset in Sabon
by codeMantra

Contents

Foreword

Dampness in the dwelling is a major reason people can be dissatisfied with what should be their home. If they are tenants, they will complain to their landlord, or if they are owner-occupiers, they will worry about what to do. There are several structural problems that can result in dampness, and the dampness can cause damage and distress, and can also have a negative impact on physical and social health.

Figure 1 Circa 1900 terraced houses.

This monograph is intended to provide a straight-forward explanation of the possible causes, and why it should be remedied, including the potential threats to health.

It is non-technical and written for every tenant and owner-occupier whose home is damp. It also aims to highlight the link between problems with the dwelling design and structure, and the interference with the occupiers' efforts to create a home.

Series Preface

This is the tenth publication in this relatively new series, and more are in the pipeline. This particular edition is aimed at a wider audience than usual and is intentionally written to be accessible to non-technical users. This is because dampness in housing concerns lawyers, community groups, and housing advisers as well as health practitioners and environmental health professionals.

It illustrates the flexibility offered by the series but the aim remains as ever; to explore environmental health topics traditional or new and raise sometimes contentious issues in more detail than might be found in the usual environmental health texts. It is a means whereby environmental health issues can be discussed with a wider audience in mind and perhaps prevent problems, as is the case with this edition.

This series is an important part of the professional landscape, as is apparent from the titles published so far and in the pipeline. Environmental health practitioners bring their expertise to a range of situations and are deployed differently but not always to the best effect so far as public health is concerned. It may be because in some countries, governments are unaware of what is environmental health, or practitioners have a 'low profile' and are taken for granted. It is hoped that this series will be used as a means of highlighting the work of environmental health practitioners.

At the same time, we want to encourage readers and practitioners, particularly those who might not have had work published previously, to submit proposals as we hope to be responsive to the needs of environmental and public health practitioners. I am particularly keen that this series is seen as an opportunity for first-time authors and as ever would urge students (whether at first- or second-degree level) to consider this an avenue for publishing findings from their research. Why for example should the hard work that has gone into a dissertation or thesis lie unread on a library shelf? We can provide advice on turning a thesis into a book.

The series provides a route for practitioners to improve the profile of the profession. EHPs have perhaps not been good at telling others about their work. To be considered a genuine profession and to develop professionally EHPs on the front line need to 'get published', writing up their work of protecting public health. This will allow them to analyse and report on what worked in practice, what was successful what wasn't and why. This can provide useful insights for others working in the field and also highlight policy issues of relevance to environmental health.

Contributing to this series should not be seen merely as an exercise in gathering CPD hours but as a useful method of reflection and an aid to career development, something that anyone who considers themselves a professional should do. I am pleased to be working with Routledge to provide this opportunity for practitioners.

As has been made clear it is not intended that this series takes a wholly 'technical' approach but provides an opportunity to consider areas of practice in a different way, for example looking at the social and political aspects of environmental health in addition to a more discursive approach on specialist areas.

Our hope remains that this is a dynamic series, providing a forum for new ideas and debate on current environmental health topics. If readers have any ideas for titles in the series please do not be afraid to submit them to me as series editor via the e-mail address below.

'Environmental health' can be taken to mean different things in different countries around the World and so we welcome suggestions from a range of professionals doing 'environmental health' work or policy development. EHPs may be a key part of the public health workforce wherever they practise, but there are also many other practitioners working to safeguard public and environmental health. So, this series will enable a wider range of practitioners and others with a professional interest to access information and also to write about issues relevant to them. The format means a relatively short production time so contents will be more immediate than in a standard textbook or reference work.

Forthcoming monographs are likely to cover such areas as Environ-mental Health in South Africa, Power-People-Planet (on leadership in the climate emergency), and Houses in Multiple Occupation. We are also in contact with colleagues around the world encouraging them to submit proposals. That does not mean we have no need of further suggestions, quite the contrary, so I hope readers with ideas for a monograph will get in touch via Ed.Needle@tandf.co.uk.

Stephen Battersby MBE PhD, FCIEH, FRSPH
Series Editor

1 Introduction

There are different causes of dampness that can affect dwellings and a series of names to describe them – penetrating, traumatic, rising, condensation, and so on. But, whatever the cause, dampness has long been recognised as an important issue, and its potential to have a negative impact on health has become an increasingly important issue.

The aim of this monograph is to summarise the structural and design matters that can lead to dampness; describe some of the structural problems that result from or are encouraged by dampness; and discuss and highlight the potential threats to health from dampness. As well as looking at the problems associated with dampness, the aim is to be practical by suggesting possible solutions. This will include looking at aspects of building design and construction that can reduce or avoid the risk of dampness, and the remedies that can solve problems.

The idea is to provide information about dampness without too much technical jargon. But, unlike many publications, it will stress and explain why dampness is a risk to the health of occupiers and highlight the need to avoid or remedy dampness to protect health.

Unfortunately, as in many areas of housing and health, there is limited up-to-date information that can be relied on by surveyors, local authority officers, building owners, and lawyers. This monograph aims to provide a resource both for those involved in the technical aspects of dampness and dwellings, and for those who want relatively non-technical information.

DOI: 10.1201/9781003080237-1

Figure 1 The 1960s purpose-built blocks of flats (single-level dwellings).

2   What Is Dampness?

There is no internationally recognised or agreed definition of what constitutes 'dampness'. So here, we explain what we mean by the term 'dampness'.

Water (moisture) is naturally present in many of the materials used in the construction of buildings. Provided that the moisture stays within certain limits (depending on the particular material), it will not cause any problems. If the moisture exceeds the upper limit for that material, problems will occur, and this is what is referred to here as 'dampness'.

Water vapour, a gas, is always present in the atmosphere. Below certain upper limits, it will cause no problems. But, above those limits, there can be problems with both the building materials and health, and these problems (threats) can occur before there are any signs of visible condensation. Above the upper limits and visible condensation are referred here as 'dampness'.

Moisture in Building Materials

Moisture is held in building materials in several ways. Water is combined with some building materials such as concrete and plaster. In the construction of a (say) two-storey, two-bedroom house, this can be in excess of 2,000 litres (depending on the form of construction). Most of this will dry out, but it may take at least 12 months, and what is left is chemically combined with the material (and will not cause any problems).

Porous building materials, including plaster, concrete, bricks, and timber, will exchange moisture with the adjacent atmosphere as a result of vapour pressure. Vapour pressure attempts to keep a balance in the moisture levels, as moisture from high-pressure areas will force moisture into low-pressure areas. Normally, this vapour pressure will

DOI: 10.1201/9781003080237-2

not disrupt the natural moisture levels of building materials, and as the porous materials are never 'truly dry' (i.e. there is always some moisture present as well as any chemically bonded water), their normal state is usually referred to as 'air-dry'.

Examples of the moisture levels for some air-dry materials in a relatively moist, but not 'damp' atmosphere, are – common bricks (not the very dense engineering bricks) will be between 1.5% and 2.5%, plaster around 1.0%, and timber around 11.0% (depending on the type of timber).

Hygroscopic Salts

Porous materials, such as brick, plaster, and concrete, can become contaminated by inorganic hygroscopic salts. These salts have an affinity for moisture and will absorb moisture from the air disrupting the balance, making the material visibly damp. There are two potential sources. One is where there is a solid fuel heating unit discharging into an unlined flue, where the flue gases cool and condense, passing salts into the chimney breast. Salt staining can also occur on the outside and inside of walls, the salts being drawn from the soil by rising dampness and left as the moisture evaporates (on which see below).

Rising Dampness

Porous materials are riddled with very fine hair-line pores, and these pores are so fine that the surface tension of water will become strong enough to draw the water upwards against gravity (the same effect as a wick or blotting paper). This effect, capillary attraction, means that materials such as bricks and concrete will 'draw' moisture out of the ground to heights of about 1 m above ground level. As noted above, the moisture can contain hygroscopic salts which, as well as the damaging effect of the moisture, will further damage the wall structure. Rising dampness can also have a dramatic effect on a solid floor that is in direct contact with soil.

It is to prevent rising dampness that damp-proof courses and damp-proof membranes are or should be incorporated into walls and floors in direct contact with the ground. In older houses (those built over 100 years old), there was often a dado, wood panelling from floor level to just less than 1 m (approx. 3 feet) above. The dado was a decorative finish that hid the damage caused by rising dampness, did not cure the problem.

Figure 2 Dampness from either penetration or a burst pipe/tank (traumatic).

Penetrating Dampness

Holes and gaps in the external fabric, the walls and roofs that should protect the interior, will allow water through that protection. Such holes and gaps may be a result of despair and lack of maintenance, or ineffective weatherproofing at the time of construction or refitting. Some gaps can be obvious, such as a slipped roof slate or tile. Often, particularly in older properties, walls will have a coating of external render (a waterproof concrete skim) to protect walls constructed of brickwork. Cracks in such render will draw water in when it rains, and as that water will not evaporate outwards, it will soak into the wall to affect the internal surfaces.

Of particular importance is the prevention of water penetration at the joints around window and door openings. These must be properly and completely sealed, and, for windows, there should be a sill at the base of the opening to throw any water running down the glazing safely away from the wall below.

Traumatic Dampness

This is when a water pipe or tank leaks or bursts, or as a result of a leak to a drainage or waste pipe serving a water-closet, sink, bath, or shower. The effect can be slow, where the leak (often at a poorly made joint) is relatively minor, but this will be enough to cause a problem over a period. Or, it can be dramatic, such as when a water tank bursts.

Problems can also occur if the water in an uninsulated pipe or water tank freezes. Frozen water (ice) expands and can break a joint or burst a pipe, and a leak resulting from this freezing will occur when the ice melts.

As water pipes and tanks are generally hidden from sight, any traumatic dampness can be difficult to trace and remedy.

Moisture in the Atmosphere

The amount of water vapour (moisture) a given volume of air can hold depends on the temperature of that air – the higher the air temperature, the more water vapour it can hold. The term 'Relative Humidity'(RH) is the ratio between the amount of water vapour held by a

Figure 3 Cracked and missing protective render to the rear main wall of a
terraced house.

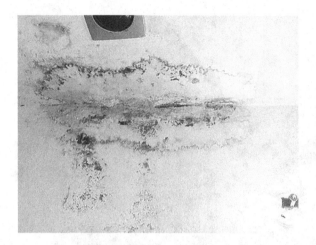

Figure 4 Damp staining and mould growth to the ceiling of upper floor room.

volume of air, and the maximum amount that volume of air is capable of holding at that temperature.

Ideally, the RH within a dwelling should be between 30% and 70%; any lower and it will start to feel uncomfortable (dry), and any higher and it will start to cause problems, such as mould growth. However, within a room (and a dwelling), there will be air temperature gradients, both horizontally and vertically; the temperature highest close to heat sources, and lowest next to cold (or cooler) surfaces. This means that the RH will differ from point to point, although the amount of water vapour may not vary.

If the RH persistently exceeds 70% then 'damp' problems will occur, although, initially, there will not be any visible signs of dampness. Visible dampness, termed condensation, occurs when the air becomes saturated (i.e. the RH reaches or exceeds 100%). This is known as the dew-point. Condensation is usually visible on cold surfaces that have reduced the air temperature, such as window glazing and cold spots on walls where heat has been transmitted from inside to the exterior.

However, condensation can also occur within the structure, and so not be visible. This happens when moist air passes through a material such as plaster (by vapour pressure) and reaches a cooler non-porous material, reducing the temperature, increasing the RH to pass beyond the dew-point. This is termed 'interstitial condensation' and may happen in timber-framed buildings (where there is a timber frame, hidden from the outside by brickwork, and from the inside by plaster board-ing), where walls have been dry-lined to improve the insulation (an internal lining of plaster board on lathes), or where there is a flat roof with no or an inadequate vapour barrier.

As well as interstitial condensation, condensation can occur when moist air from the living areas is able to reach the roof space (loft) under a pitched roof, and in the sub-floor space (the space below a suspended timber floor).

The production of water vapour in an occupied dwelling is a nat-ural result of domestic and biological functions and activities. There are various estimates of the amount of moisture generated by a four-person household. A realistic estimate is that such a household can generate 30–40 litres of moisture per week just by breathing, add to that 15–20 litres a week by cooking, showering, and bathing. If laundry is dried indoors (in inclement weather), this will add a further 35 litres a week. This means that such a four-person household will generate anywhere between 45 and 95 litres of water vapour a week purely from normal domestic and biological functions and activities.

Estimated range of moisture emissions for a four-person household.

Source	Emissions over 24 hours (litres)
Everyday sources	
4 persons asleep for 8 hours	1.0–2.0
2 persons active for 16 hours	1.5–3.0
Cooking for 3 hours (by gas)	2.0–4.0
Bathing, dishwashing, etc.	0.5–1.0
Basic Regular Total	5.0–10.0
Irregular sources	
Washing clothes	0.5–1.0
Drying clothes (indoors during inclement weather)	3.0–7.5
Flueless gas or oil heater (4 kW for 5 hours)	1.0–2.0
Possible daily total	10.0–20.0

The amounts emitted will depend on the size and composition of the household, and the amount of time spent in the dwelling; for example, a household of two older people would spend more time indoors than a household consisting of a working couple. A household including young children may also spend a lot of time indoors. While a dwelling should be of a size and layout to suit the possible occupying household, this will not always be the case as occupation changes over the years (households moving, children being born, and later moving out). This means that dwellings should be capable of being occupied by a spectrum of households without any condensation problems.

There will be variations between different households, but these will be surprisingly small. Even a single-person household will generate between 4.5 and 10.0 litres over 24 hours.

While the normal day-to-day activities of occupiers will have some effect on the amount of moisture generated within the dwelling, this is unlikely to be a major contributor to high RH or condensation (see the Table in the preceding section). Unfortunately, it is all too easy for building managers and landlords of rented dwellings to lay the blame at the door of the occupiers, rather than solve why the dwelling does not meet the basic principle of providing living accommodation able to cope with occupiers. Crowding can mean excess moisture, but this is a different problem. Expecting occupiers to change their lifestyle to reduce moisture production is expecting occupiers to compensate for problems attributable to defects including lack of maintenance poor design and/or construction of the dwelling.

Flooding

Two factors have increased the likelihood of dwellings being flooded. One is a result of changes in weather patterns that now mean there are more heavy rain storms and flash floods, another factor is the greater use of potential flood plains for housing developments.

UK houses were not, and are not yet, designed to avoid flood water entering. Flooding, even by a small amount, has a dramatic impact and requires extensive remedial action after the water has subsided, particularly as flood water will be contaminated, usually by sewage.

3 Sources of Dampness and the Potential Effect on the Structure

There follows a brief summary of potential sources of dampness and the possible consequential effects. This summary is not intended to be comprehensive, and full and more detailed information and consideration can be found in publications on construction and the functions of structural elements.

Roofs

Pitched roofs are intended to direct water safely away to eavesgutters and rainwater goods. Pitched roofs are usually finished with tiles or slates (although some may be finished with metal, such as copper, or bitumastic felt).

The weakest point in any roof is where it is penetrated, such as by chimneys, pipes, and dormer windows. The weak point is the junction between the roof covering (tiles or slates) and the chimney, pipe, or dormer. Chimneys in older houses are usually of brickwork, and cement or a malleable material (such as lead) is used to provide a water-proof seal, termed 'flashing'. This flashing, or the adjacent brickwork, may deteriorate allowing water to enter the roof space and dwelling, often causing dampness to the plastered surface of the chimney breast within the living area.

So-called flat roofs should be designed to prevent water collecting (ponding). It is often difficult to see a flat roof easily, and this may mean that any ponding and its damage may go undetected for some time. Often, the first indication will be damp stains and/or penetration affecting the ceiling below the damage.

As well as the potential effects on the living area, slipped slates or tiles, or damaged covering will allow water penetration into the roof

DOI: 10.1201/9781003080237-3

Figure 5 Slipped and missing slate allowing penetration. (Also shows cement flashing between the chimney stack and the roof slates.)

structure, and that water may damage roof timbers or ceiling plaster to upper floors, such damage going unnoticed.

Rainwater Goods

These include eavesgutters intended to safely collect rainwater from roofs, and downpipes to collect the water from eavesgutters and take it safely away from the building. If these become blocked, water will overflow and can damage the adjacent wall, washing out the mortar of brickwork joints and the surface of bricks. The amount of water may result in penetration through the wall into the interior.

Where it is likely for snow to collect on roofs, it can slip as it starts to melt, and can result in too much water for the eavesgutters to cope, or the weight may over-power the eavesgutters causing them to overflow or even collapse under the weight. Where snow is regularly expected in winter periods, guarding should be fitted to retain the snow until it melts safely. Rainwater goods are vulnerable to blocking caused by freezing during cold periods. Ice expands as water freezes and can cause damage to downpipes.

Figure 6 Staining to the wall indicating overflowing rainwater fallpipe. As well as damaging the brickwork, this could result in dampness inside.

Brickwork

Moisture in the soil will contain soluble salts, including salts of nitrate and chloride. These will be held in solution and, through rising dampness, into the structure of walls and solid floor. As the moisture evaporates, the salts will be left behind. Although the amounts are relatively minute, the salts deposited, over many years, will have some effects.

Brickwork in older houses without a damp-proof course will suffer with rising dampness. The salts deposited in the brickwork will initially leave a stain at the height reached by the moisture. This stain is, perhaps unsightly, but, in itself, not really a structural but a decorative problem. However, the dampness will damage the integrity of the bricks, and the outer surface will, eventually, loosen and spall (detach and fall-off). The salts in moisture from the soil are hygroscopic, and absorb moisture from the air, and can, ultimately, have damaging effects.

Door and Window Openings

These are weak points in any external walls. The joints between the frames should be water-tight to prevent penetration. The sill to windows should slope away from the wall below and should incorporate a groove on the underside so that any water running off the window and frame drips away from the wall.

There should be a weather bar to the base of external doors (or on the threshold where the closed door fits) to prevent the penetration of water between the door and threshold.

There should be drip grooves to the head of door and window openings to divert water running down the wall away from the door and window openings.

External Drainage and Waste Pipes

These are the pipes taking soil and waste water from facilities (such as baths, showers, wash hand basins, and WCs) from within the dwelling and discharging it into the drains and sewers, or sumps or septic tanks. Such pipework can become damaged or blocked, allowing the contents to spill onto adjacent wall surfaces, again, damaging the structure. The water in these external pipes can freeze in extreme cold weather. Waste water is less likely to suffer because of its temperature, but it is still possible. As ice expands, it can damage joints.

Solid Floors

Similar to brickwork in older houses, solid (concrete) floors without a damp-proof membrane will be affected by rising damp. Most such floors will have a finishing surface on top of the concrete. While relatively protected by the finishing surface from drying out and evaporation, the dampness will damage the material of the joint between the finish and the concrete so that the floor finish will become detached and loose.

Reinforced Concrete

Unprotected ferrous metals will rust (a reddish brown oxide of iron) when exposed to water and air. As concrete does not provide any protection in itself, rusting of the reinforcements can be a serious problem; and, as the metal rusts it expands and the slow force generated can shatter the concrete.

Plasterwork

Rising dampness will affect internal plasterwork in older houses without an effective damp-proof course. Gradually, the affected plaster will deteriorate and become detached from the underlying wall (this effect is termed 'blown').

Internal plaster will also be affected by condensation, but this is less destructive to the structure, but will damage decorations and attract various forms of mould growth.

Ceilings (and walls) will be damaged by dampness penetrating through disrepair to roofs, by burst pipes, and leaking joints to sanitary ware, personal washing facilities, and storage tanks (in lofts). Penetration through roofs will be associated with rainfall. Traumatic damp associated with facilities will be linked to their use; although this may be sudden and dramatic, and the weight of the water may cause the plaster to collapse.

Figure 7 Plaster affected by penetrating dampness.

Timber

Not all timber used in construction is visible, being an internal part of the structure. For example, 'hidden' timbers can include floor joists supporting suspended ground floors, joists between ceilings and upper floors, joists supporting the ceiling to top floors, roof rafters, and timber within the structure of walls.

Unprotected timber (particularly soft wood) used for windows and doors and their frames will be intermittently affected by dampness (such as rain and snow) particularly when protective paint is cracked and flaking. This will cause warping and distortion. This can also be caused by a constantly damp environment. The result is that the windows and doors will stick and may not close and open easily or properly.

Damp-affected timber, with a moisture content of 18% or more, will be vulnerable to fungal attack. This is especially so for structural timbers (joists) to suspended ground floors where there is poor and inadequate under-floor ventilation. The most serious of timber attacking fungi is generally known as 'dry rot' (*Serpula lacrymans*). It produces a mass of mycelial (hair-like filaments) that spread through the timber, extruding enzymes which digest and weaken the timber. Dry rot is particularly virulent and can spread through brickwork and concrete. Although considered less serious but of importance is the so-called 'wet rot' (*Choanephora cucurbitarum*). As well as damaging the structural integrity of timber, dry rot and other fungi produce spores, typically single-celled reproductive units. Such spores spread the fungus, but also pose serious threats to health.

Damp timber may also attract termites which can cause damage and weaken the timber (although termites are rare in the UK, but widespread in Europe and the USA).

Energy Efficiency

There are several factors that relate to the energy efficiency of a dwelling:

- Thermal insulation of external walls, of windows, of roofs (or the ceiling to upper rooms), and of floors.
- Provision for space heating.
- Provision for ventilation.

Thermal Insulation

Older dwellings, constructed to meet earlier standards, are usually energy inefficient. The main structural factor is the thermal insulation provided by the structure (the fabric of the building). Brick-built walls were traditionally solid, and usually of 23 cm (9 inch) thickness. Such walls provide poor thermal insulation and poor protection from heavy rain. To improve protection from rain penetration, cavity walls were introduced; these had a inner skin of one brick thickness (about 11.5 cm, 4.5 inches) a gap (of around 5 cm, 2 inches) and an outer skin of one brick thickness. An unintentional gain was slightly improved thermal insulation provided by the air gap between the skins. Subsequently, requirements were introduced to increase the thermal insulation provided by external walls.

One method to increase the insulation given by external walls was the injection into the cavity of material such as polystyrene foam. Some forms of insulation to external walls can settle, reducing the insulation to upper sections of walls, and some can be affected by moisture, either from rising dampness or rain penetration. As water is a 'good' conductor of heat, damp-affected insulation will be relatively ineffective, resulting in loss of energy and heat.

Windows in older dwellings often consist of a single pane of glass (unless up-graded) and will be major heat loss components. Properly formed double or even triple glazing will reduce heat loss considerably, and so reduce the risk of condensation. Where double or triple glazing is not an option because of the type of window or planning restrictions, secondary glazing can help.

The ceilings to top floor rooms should be divided from the roof space (or the outer surface of flat roofs) with insulation (and a barrier to prevent water vapour passing into the roof space). Alternatively, the underside of the waterproofing outer surface of a pitched roof should be insulated; providing a useable loft space.

Suspended timber ground floors can lose heat into the under-floor space, and solid floors can transfer heat to the ground. In both cases, there should be a vapour barrier or damp-proof membrane, and, in the case of solid floors, and insulating layer.

Space Heating

The provision for space heating should be designed to distribute heat throughout the living area of the dwelling, be controllable and be affordable. The type of heating should take account of the construction of the dwelling. Walls constructed of dense material will be slow to heat up and cool down, and will give off heat over a period. Heating systems that provide relatively instant heat, such as radiant and warm air systems, will heat the individuals quickly, but will be (very) slow to heat dense structure. Lightweight inner surfaces will heat up quickly, but will not hold that heat.

Water heating should also be affordable, and should be capable of heating the water to around 16°C; any hotter could cause scalding, and cooler may allow bacteria (such as *Legionella*) to multiply. Where a part of the provision for hot water includes storage tanks, these should be insulated to prevent heat loss from the water and so save energy. Cold water tanks and pipes should be insulated to protect against freezing.

Ideally, flueless gas or oil heaters should not be used in dwellings (although they are used where, for economic reasons, the occupying household has been disconnected from a mains gas or electricity supply and left with no alternative for space heating). The fuel for such heaters is not economic, and the heaters emit all the products from combustion as well as heat into the atmosphere within the dwelling (see Table "Estimated range of moisture emissions for a four person household").

Ventilation

This is necessary to replenish the internal air and remove excess moisture. Means of ventilation should be controllable and not excessive, and should help ensure relative humidity (RH) is kept at safe levels and heat loss is kept to a minimum.

In general, there should be around 0.5 air changes per hour, with increased localised extraction in specific areas, the kitchen, and bathroom, during periods of high moisture production (such as cooking and showering). Air changes above 0.5 for general parts of the dwelling will increase the heat loss. Windows do not need to be open wide, and most modern double-glazed units incorporate trickle vents which should be sufficient in most rooms.

For the kitchen and bathroom, there should be extractor fans located and designed to take moisture-laden air out of the dwelling. Such fans can be fitted with humidity-sensitive switches, to automatically operate when moisture levels are high, and also with a manual control.

Structural damage associated with energy inefficiency as a result of an imbalance in the factors will be primarily high RH (i.e. above 70%) or condensation (visible moisture). These will enable mould and/or fungal spores (always present in the atmosphere) to germinate on surfaces. Where mould appears on wall or ceiling surfaces it will not cause damage to the material, although it will damage decoration. The presence of high RH or condensation on timber will enable the spores to germinate and attack/infect the timber, ultimately weakening it.

4 Health Effects of Dampness

High relative humidity (RH) levels (those above 70%) can threaten health in several ways:

- It can reduce the ambient air temperature and so affect thermal comfort.
- It can interfere with the body's natural cooling mechanism, which relies on evaporation.
- High RH and the general presence of dampness will cause damp clothing and bedding, which will feel cold as moisture evaporates, meaning that occupiers are likely to use more heat in order to feel warm.

The health effects of indoor microbial growth – especially fungal growth – have been studied extensively for over 20 years. The studies have provided a wealth of reports and material on the health effects of dampness and the associated consequences, but the most serious threats to health are linked to the development of microorganisms. These are ubiquitous and can propagate rapidly wherever moisture is available, and even the small amounts of dust and dirt normally present in most indoor spaces will provide sufficient nutrients to support their growth. Besides excess moisture, the development of microorganisms, such as moulds, is affected by temperature, ventilation, and building materials.

As stated in the preceding section, the normal day-to-day activities of occupiers will have some effect on the amount of moisture generated within the dwelling, but will be minor and not likely to be a contributor to high RH or condensation (see the Table "Estimated range of moisture emissions for a four person household"), and occupiers should not be blamed for condensation or high RH – dwellings should be designed for occupation.

DOI: 10.1201/9781003080237-4

Mould and Fungal Spores

The word mould refers to fungi, most of which can grow on building materials. There is a very wide variety of fungal species – probably several million – that thrive under different conditions. Moulds produce reproductive structures such as spores which, when released, are mostly suspended in the air and disperse to spread the moulds to other areas. Moulds are also able to synthesise chemicals (mycotoxins, microbial volatile organic compounds) that are contained in the spores or released directly into the air.

Damp-affected timber is vulnerable to fungal attack and infection. As spores of a wide range of fungi and moulds are always present in the atmosphere, damp timber provides an ideal medium for their germination and enables the fungus to become established. Wet rot (*Choanephora cucurbitarum*) and dry rot (*Serpula lacrymans*) are virulent timber attacking fungi and will dramatically increase the spore content of the atmosphere within a dwelling, even though the source, the fungus, may be hidden.

Other sources of spores are moulds infecting damp plaster and other surfaces (both visible and hidden surfaces). These spores are encouraged by the moisture naturally in the material and by any condensation (visible moisture) or high levels of water vapour because of high RH.

Indoor mould growth is considered by public health organisations (such as the World Health Organization (WHO)) as a significant health hazard. In 2004, the Institute of Medicine (IoM) conducted a comprehensive review of the scientific literature into the relationship between damp or mouldy indoor environments and the occurrence of adverse health effects, particularly respiratory and allergic symptoms (Institute of Medicine 2004). In 2009, the WHO published an extensive review and guidelines on dampness and moulds (World Health Organization 2009). These reviews are still considered as references as more recent epidemiological data mostly confirm their conclusions.

Damp conditions and mould growth in homes increase the risk of respiratory allergy symptoms and exacerbate asthma in sensitive individuals. People living in homes with mould and dampness are more likely to have symptoms such as eye, nose, and throat irritations, coughing, wheezing and shortness of breath, and the worsening of asthma symptoms. The level of concern depends on the extent of mould, how long it has been present, and the sensitivity and health status of the residents. Most susceptible are infants, children, the elderly, and those with health problems, in particular those with respiratory conditions such as asthma allergies, or chronic obstructive lung conditions, and weakened immune systems.

High levels of mould and fungal spores can be allergenic, and exposure over a period can sensitise atopic individuals (those with a predetermined inherited tendency to sensitisation) and may even sensitise non-atopic individuals. Once sensitised, exposure to even a relatively low concentration of the spores can trigger allergic symptoms such as rhinitis, conjunctivitis, eczema, cough, and wheeze and can lead to and exacerbate asthma.

As well as the spores of moulds and fungi aggravating existing respiratory conditions, the spores of some moulds can be aggressive. For example, some can be carcinogenic, others toxic, and some infectious. Some moulds produce toxins (mycotoxins) that can cause nausea and diarrhoea. Studies have also described the involvement of mould sensitisation in the severity of inflammatory and allergic pulmonary responses.

Most published studies on the health effects involve qualitative assessments of dampness or mould, including visible water damage, visible moisture, leaks, flooding, visible condensation on windows, visible mould or mildew, and mouldy or musty odour. Fewer findings are available on quantitatively measured microbiologic factors. Also, studies presenting a quantitative characterisation of the exposures are heterogeneous (diversity of measurement methods) and make it impossible to give a health threshold.

Figure 8 Damp and mould growth in a child's bedroom.

Nonetheless, studies have shown that the observed presence of mould in living rooms and the musty smell, whether or not studied in conjunction with dampness, are associated with the development of asthma in young children with robust evidence suggesting a causal relationship. They have also shown an association between exposure to visible moulds and the risk of allergic rhinitis, but further longitudinal studies are needed to confirm a causal relationship.

House Dust Mites

Damp houses significantly increase exposure to dust mite allergens, at least in populations living in mild and temperate climates (World Health Organization 2009).

House dust contains several species of mites. Of particular relevance are the house dust mite (*Dermatophagoides pteronyssinus*) and dust mite (*Glycyphagus domesticus*). These have been shown to be potent allergens, not only in themselves, but also in their excretory and secretory products (including cast skins, faeces, and carcases). Both of these mites thrive in warm moist atmospheres, the optimum temperature being around 23–25°C (temperatures also favoured by some households). While the populations remain relatively stable where the RH is 60% or below, the populations will dramatically increase where the RH is above 80%.

These mites, virtually invisible to the naked eye, are found in all dwellings, in the dust, in mattresses and bedding, and in soft furnishings. Mites feed on human and animal dander, the dead skin dust flaking off occupiers and their pets. This dander collects in bedding and soft furnishings, and it seems that while anti-mite and mite control measures, such as vacuuming beds and furnishings, have minimal effect, the most successful action appears to be the reduction in dampness (both visible and high RH levels).

Other Health Threats

In addition, although pests are not specific to damp buildings, standing water may attract and support cockroaches and rodents infestations, which can transmit infectious diseases and are a source of allergens.

Excessive moisture may also enhance emissions of chemical substances from building materials and furnishings that will alter indoor air quality and consequently have an impact on health and well-being.

As well as the potential effects on physical health, damp and cold ambient temperatures, and the visual signs of mould can affect mental and social health by making adult and young occupiers reluctant to invite friends into the dwelling. Adults will be affected by stress and anxiety, and adults and children will become isolated from friends and the community.

It has also been suggested by some studies that exposure to damp buildings could increase the risk of sick building syndrome, but a causal relationship has not been demonstrated.

5 Identifying and Remedying the Cause(s) to Protect Health

Ultimately, in order to protect health and avoid (further) damage to the structure, it is necessary to remedy the causes and sources of dampness and prevent its recurrence. The threats to the health of the occupiers cannot be removed and recurrence prevented unless the cause/source of the dampness is resolved. Once the problem of dampness is remedied, then the threats to health must be safely removed. However, dealing with the threats to health may be necessary in the short term while the remedial measures are completed.

Here, we consider first dampness relating to structural problems, then dampness relating to energy inefficiency, which can be complex, and finally dampness following flooding.

The causes of dampness result from various factors, including:

- Inappropriate construction materials.
- Poor or inappropriate design.
- Poor or incompetent workmanship in the construction or alteration.
- A lack of proper and thorough repair and maintenance.
- Energy inefficiency (including thermal insulation, space heating, and ventilation).
- Flooding.

Identifying the Source of the Dampness

First, there should be a general inspection of the property (on which see Annex). If the dwelling is owner-occupied this will probably be carried out by the occupier, but a full inspection and investigation made by an experienced professional may be necessary.

DOI: 10.1201/9781003080237-5

If the dwelling is rented (either from a private or a public land-lord) it is in that landlord's interest to protect their property (their investment) as well as protect the tenant's health, by ensuring that any dampness is remedied.

The inspection should begin with the exterior, covering the potential weak areas as mentioned in the section on Causes and Structural Effects, focusing on the physical structure and fabric of the property. Each element of the fabric structure (and of the dwelling as a whole) should serve its intended purpose. Any apparent design faults or evidence of poor or inadequate workmanship or maintenance should be remedied.

Each individual element has a 'lifespan', some much shorter than others. This means that some elements will need replacement and renewal before others. This is often taken as an opportunity to upgrade an element, but by doing this, the implications and possible effect on the dwelling should be taken into account. A dwelling is a dynamic unit, and changing one element or system may affect another. For example, the replacement of single-glazed windows with double-glazed ones may reduce the heat loss through the window areas, but may interfere with the ventilation; and upgrading or changing the heating system may have similar effects.

For a traumatic problem, if major or sudden, it is usually relatively easy to identify the source. Comparatively more problematic may be slow leaks, where the water 'runs' down internal elements (this also applies to problems of slow penetration through the roof, the water running down rafters). Nonetheless, the source of the dampness must be traced and rectified.

Dampness Related to the Design, Construction, or Repair

Where the cause of the problem is attributable to inappropriate materials, incompetent workmanship, a lack of repair and maintenance, or inappropriate design, the solution will involve replacement and may even involve reconstruction.

Determining the cause and source of dampness attributable to these matters will involve an inspection of the dwelling. As mentioned above, this can be carried out by the occupier, but ideally an inspection should be carried out by someone with specialist knowledge. Whoever undertakes the inspection should at the least follow the process outlined in the Annex.

Energy Efficiency

One of the most difficult problems influencing the likelihood of dampness is the thermal/energy efficiency of the dwelling, particularly older dwellings. Determining the most appropriate solution is a specialist task.

As stated above, it is important to note that dwellings are, or should be, designed and constructed for the purpose of occupation, and that means that they should be capable of handling the moisture produced by normal biological and domestic activities without problems occurring, such as condensation and/or mould growth. However, that capability may be compromised when alterations (including perceived 'upgrading' or 'modernising') are carried out. In addition, over the years, there will probably be changes to bathing/showering practices; to clothes laundering and drying; to heating, and to the type of energy used (electricity, gas, or solid fuel). These changes will all affect the amount of moisture produced within a dwelling, the ventilation, and how the dwelling copes with moisture generally.

Additionally, the changes in climate, such as extreme weather events, may also affect whether a dwelling is able to cope with the moisture generated without adaptations.

Many of the fundamental biological and domestic activities and functions will generate moisture. The amounts emitted will depend on the size and composition of the household, and the amount of time spent in the dwelling (on this see above) For example, a household of two elderly individuals or a household with small children would spend more time indoors than a household consisting of a working couple. As changes in the occupation are likely over the years, a dwelling should be designed to be capable of being occupied by a spectrum of households.

Remedying condensation problems involves ensuring there is the right balance between thermal insulation, space heating, and ventilation. As already stated, this is a specialist task.

Mould Growth

If it is not possible or practicable to remedy the main underlying structural problem immediately, and if the dwelling is occupied, it will be necessary to deal with the immediate threats to health, such as mould. All areas should be checked for mould growth, including 'dead' areas (areas where there is little air movement, stagnant areas) such as surfaces hidden normally by furniture (e.g. wardrobes).

While small patches of mould can be treated by the occupier, ideally, mould removal should be carried out by a specialist. Whatever the option, there are important considerations to take into account, these include:

- The size of the infected area. If this is extensive or in several areas, the area(s) should be cleaned and treated.
- Whoever tackles the problem should wear overalls (or similar protective clothing) and a mask. If mould is disturbed, it will release spores, so as well as the protective clothing and mask, the room or area should be well-ventilated.
- Children, elderly, immuno-compromised, allergic, and asthmatic individuals should not be present within the dwelling until the cleansing is safely completed.

While not usually necessary, if it is considered needed or appropriate for the mould species to be identified, this will involve sampling. This should be carried out by a trained person.

Figure 9 A 1930s detached house.

Annex

Investigating the Cause/Sources of Dampness

This is a brief outline of the approach to investigating dampness, and it is followed by a checklist to help with that investigation.

Signs of potential sources of penetration of dampness through the roof(s) include slipped or missing slates, cracked tiles. A full examination of roofs including flashings and valleys will not be possible without ladders. Similarly, the condition of flats roofs may not be possible without means of access (or a view from an adjacent higher building!).

Dampness from these sources in particular will reflect the weather – the damp internally will be affected by rainfall, and will be less in drier periods.

Blocked, cracked, or leaking rain water goods will be apparent by the staining to adjacent wall surfaces. Water from defective eavesgutters and down pipes will erode render and brickwork (particularly the mortar joints to a brick wall). Again, the weather will have some effect on the internal dampness.

Rotting timber to window and door frames may be a source of internal dampness, as may poor or inadequate sealing.

Internally, the dampness or staining may be patchy, usually coinciding with the external defect.

Where soil or paving covers the damp proof course (dpc), this will negate the dpc and allow rising dampness. A dpc should be at least 150 mm (6 inches) above the adjacent soil or paving level to avoid rain bouncing above the dpc. Internally, rising dampness will create a 'tide' mark to the plaster to a height of about 1 m (3 feet). Mould to rising damp internally is rare as the moisture will contain salts drawn up from the soil which will inhibit mould growth.

Damp patches can occur on chimney breasts where this is or was an open fireplace, and sometimes where there is a gas fire or a solid fuel stove. These result from hygroscopic salt deposits (salts that absorb moisture from the air), and where the flue has not been lined. This occurs when the products of combustion within the flue cool and the resulting condensate contain the salts which are deposited within the unlined walls of the flue. As they build up, they begin to absorb more moisture, damaging the plaster and sometimes leaving a brownish stain. Lining the flue with a waterproof material or liner prevents this problem.

The presence of mould growth with no obvious external source of moisture, usually indicates that the problem is condensation. This affects areas that are colder than others. For example, areas above window and door openings where a dense material is slower to warm-up than adjacent areas; where there is a link across the cavity, either by mortar on wall ties or lintels above openings[1]; or at corners of walls where heat loss is higher than elsewhere. Mould growth is more likely in cooler and perhaps unheated, rooms, and areas.

A visual and superficial inspection may not give indications as to the cause, but may help to eliminate the least likely ones. However, a detailed inspection by a specialist may be necessary, and, in some cases, may involve opening up the structure.

Dampness Checklist

Fully and properly assessing the cause and identifying the most appropriate solution should be carried out by a specialist – such as a surveyor, a heating engineer, or an environmental health practitioner. But here we give a basic checklist that can help understand what could be the cause and the possible remedies.

The first stage is to carry out a thorough visible check internally to look for the characteristics, the signs of the problem, using the checklist below. The code given refers to the probable cause of the dampness, and the next stage in the assessment. (For an explanation of the causes, see Chapter 2 and a brief outline is given in the table below.)

Code	Cause of dampness
C	Condensation
P	Penetrating
R	Rising
T	Traumatic

1. Check internally and tick the visible sign(s).

Signs	Code
Staining to the ceiling of upper floor room(s).	P/T
Staining to external wall(s) around windows or doors.	P
Staining in patch(es) to external wall(s).	P
Staining to the ceiling of lower floor room.	T
Water leaking from joint (or crack) to waste, soil, or water pipe (i.e. below sink or wash hand basin, or to pipe from WC basin).	T
Staining to the base of external ground floor wall(s) to height of about 1 m (3 feet).	R
Rotting skirting boarding to external ground floor wall(s).	R
Wallpaper (or paint) peeling off at base of external ground floor wall(s).	R
Floor covering to ground floor room(s) lifting.	R
Misting of windows.	C
Staining above window or door opening(s).	C
Mould growth to external wall(s) and/or ceiling(s).	C
Water collecting on internal window sills.	C
Staining in patch(es) at the junction of external walls.	C
Moisture on the surfaces of tiles (or other hard surfaces).	C
Peeling wallpaper generally.	C
Damp or mouldy clothes.	C
Mould behind furniture (such as wardrobes, chests of drawers, and sofas) or equipment (fridges and freezers).	C

2. The next stage in trying to find the cause is to look for the possible problems.

Code	Probable cause of the dampness
C	**Condensation.** This is when warm moisture-laden air is cooled by a relatively cold surface. The moisture will be at its worst during the colder periods of the year, and mould is most likely for this form of dampness.
P	**Penetrating.** Rain water penetrating through the external surface(s) of the dwelling. This form of dampness will be worse when there is rain or it has been raining. It may also occur when snow melts on a roof or in eavesgutters.
R	**Rising.** Where moisture from the soil rises up the wall or floor because of capillary attraction. This form of dampness will not be affected by rain and will be fairly constant throughout the year, only being drier during long spells of hot weather.
T	**Traumatic.** This is the result of a leak from, or bust to, a water, drainage, or sewage pipe, or a water storage tank. If the problem is to a ceiling or internal wall other than those to the top floor, then the dampness may be noticeably worse when a particular facility is used.

3. Next, check for other signs that suggest what might be the possible cause.

Code	Signs	Comment
C	The dampness will not be affected by rain, but may be most noticeable on the inside of the most exposed walls.	This is most likely to occur where the heating system doesn't extend to all parts of the dwelling, where there are problems with the heating system (such as faulty controls or disconnection), or where portable flueless gas or oil heaters are used.
P	The site of the internal damp will give an indication of where the outside problem is likely to be – e.g. if the staining is to a ceiling to the top floor room, then the problem is likely to be the roof; if the staining is to a wall then it could be damage to the wall such as cracked or missing render, defective rainwater goods. For more details and examples, see below.	
	Slipped, cracked, or missing slate or tile to a pitched roof. Or, cracked or deteriorated flashing between the roof and chimney stack or vent pipe penetrating the roof.	This could allow rain water (or water from melting snow) to penetrate into the roof space and cause dampness to a ceiling to an upper floor room or landing.
	(NB: Usually, it will not be possible to check the state of a flat roof.)	
	Leaking rainwater eavesgutter, downpipe, or soil pipe.	There may be staining around the leak (most probably at a joint) and staining to the wall adjacent or below to the leak.
	Cracked or missing render.	This will allow moisture to penetrate into and through the wall.

(Continued)

Code	Signs	Comment
	Loose or missing external sealing around a door or window frame and the opening.	This would allow moisture to by-pass the frame and cause dampness to the plaster around the reveals of the door or window opening.
	Peeling paintwork and/or rotting timber to window or door frames.	This will allow water to pass through into the interior (if not immediately, eventually).
R	Staining to the lowest surfaces of external walls.	There may show similar staining to the outside as that appearing inside – staining to a height of about 1 m (3 feet), perhaps with a 'tide-mark' at top.
T	Staining to ceilings and/or walls internally, and this may be worse when a particular facility is used.	In this case, the problem will be from an internal source. The pipes to and from facilities should be checked, including wash basins, baths, showers, WC cisterns and soil pipes. If the staining is to the ceiling to the top floor, this may mean a leak or burst to a tank or pipe in the roof space.

4. What needs to be done?

Here, we give, briefly, the possible remedies.

Code	Possible solution(s)

C   Oil or gas flueless heaters should not be used (except in an emergency such as a power cut) as these put water vapour (and other gases) into the atmosphere. The same applies to gas cookers (hob and oven), these should not be used for heating.

The space heating system should be thoroughly checked for any defects, and to see if it can be improved to provide more heat. It may be necessary to replace the system with one more efficient and one providing more heat output to the whole of the dwelling.

The best option for assessing the heating system would be to involve a Heating Engineer to determine and advise of the most appropriate solutions.

If the windows are single glazed (only one pane of glass), these should be replaced with double (or even triple) glazed units. Importantly, such units should incorporate trickle ventilators.

Fitting extractor fans (with humidity sensitive switches) to bathrooms and kitchens can take moisture-laden air outside helping to reduce the total amount of water vapour within the dwelling.

If possible, extra insulation should be added to the external walls. If the walls are of cavity construction (i.e. constructed with two leaves of brick or stone work, with a gap between the leaves) then insulation should be inserted to fill the cavity. Adding insulation to a solid wall is problematic – ideally, it should be added to the outer face (although this may not be possible if the dwelling abuts a footpath or public area); if added to the internal face then it will reduce the size of the rooms (albeit slightly, by say 50 cm – 2 inches to the inner surface of external walls).

P   Any defects/disrepair to the exterior should be remedied – the structure should be weather-proof and provide protection for those within the dwelling.

Slipped, cracked or missing slates or tiles must be replaced and properly fixed. And any flashing sealing the junction between the roof and a chimney stack, vent pipe, or wall should be checked, and, if necessary, replaced.

The seal between window and door frames and the surrounding reveals must be sound, and if missing or cracked replaced with an appropriate sealant.

Any rotting woodwork to window and/or door frames must be replaced, and all timber protected with an appropriate external paint.

(Continued)

Code Possible solution(s)

Rainwater goods (eavesgutters and down pipes) should be checked for blockages. While it may be possible to carry out a temporary repair of a leak to a gutter, a downpipe or soil pipe, it will ultimately have to be properly fixed or replaced.

Render acts as a waterproof protective finish, and when it is cracked or a piece is missing, water will get between the render and the wall structure, loosening the render and making it ineffective. This means that the area around a crack or missing area must be removed until secure render is found, and then the exposed area re-rendered. Alternatively (and probably better) would be to remove and replace the whole of the render to the affected wall.

R   If the damp-stained external area of the wall has not deteriorated (e.g. flaking bricks or missing mortar joints), then there will be no need to take any remedial action to the affected wall structure. If there is any deterioration, this should be made good – flaking bricks cut out and renewed, and mortar joints renewed.

To stop any heavy rain 'bouncing' up onto the wall, an existing damp proof course (dpc) must be at least 15 cm (6 inches) above the adjacent ground surface (whether it is soil or a hard surface). If the dpc is less than 15 cm. then the adjacent ground (whatever it is) must be lowered. If necessary, this can be a by means of a trench at least 30 cm (1 foot) wide.

If there is render extending over a dpc this must be removed as it will provide a bridge for moisture to by-pass the dpc.

Where there is no sign of a damp proof course, then a new one must be inserted. This is specialist work and may involve injecting a chemical into the wall to give a waterproof barrier, or the wall cut at a mortar joint and a bitumastic dpc inserted.

T   Any leaking pipe or tank must be removed and properly replaced. This should be done by a specialist (a plumber).

References and Further Reading

Building Research Establishment (2004). Understanding dampness – effects, causes, diagnosis and remedies. Available at – https://www.brebookshop.com/details.jsp?id=148923

Center for Disease Control and Prevention (2020). Basic facts about mold and dampness. Available at – https://www.cdc.gov/mold/faqs.htm

Housing Ombudsman (2021). Spotlight on: Damp and mould – It's not lifestyle. Available at – https://www.housing-ombudsman.org.uk/wp-content/uploads/2021/10/Spotlight-report-Damp-and-mould-final.pdf

Institute of Medicine (2004). Damp indoor spaces and health. Available at – https://www.nap.edu/catalog/11011/damp-indoor-spaces-and-health

National Health Service (2018). Can damp and mould affect your health. Available at – https://www.nhs.uk/common-health-questions/lifestyle/can-damp-and-mould-affect-my-health/

World Health Organization (2009). WHO guidelines for indoor air quality: dampness and mould. Available at – https://www.euro.who.int/__data/assets/pdf_file/0017/43325/E92645.pdf

World Health Organization (2009). DAMP AND MOULD: Health risks, prevention and remedial actions. (Information Brochure). Available at – https://www.euro.who.int/__data/assets/pdf_file/0003/78636/Damp_Mould_Brochure.pdf

Index

Note: *Italic* page numbers refer to figures

Printed in the United States
by Baker & Taylor Publisher Services